A CALL FOR LEADERSHIP

Accessible, abundant, and affordable electric power is one of the cornerstones of the U.S. economy. On February 6, 2003 President Bush highlighted the need to "…modernize our electric delivery system…for economic security…and for national security." The U.S. Department of Energy is committed to leading a national effort to do this.

The National Academy of Engineering has called the North American power grid the "…supreme engineering achievement of the 20th century." One of the aims of this document is to envision a future electric system for North America that will be considered the supreme engineering achievement of the 21st century.

Modernizing America's electric system is a substantial undertaking. The Nation's aging electro-mechanical electric grid cannot keep pace with innovations in the digital information and telecommunications network. Power outages and power quality disturbances cost the economy billions of dollars annually. America needs an electric superhighway to support our information superhighway.

Change of this magnitude requires unprecedented levels of cooperation among the electric power industry's many stakeholders. Hundreds of billions of dollars of investment will be needed over the coming decades to accomplish modernization of the electric system. National leadership is needed to create a shared vision of the future and to build effective public-private partnerships for getting there.

Imagine the possibilities: electricity and information flowing together in real time, near-zero economic losses from outages and power quality disturbances, a wider array of customized energy choices, suppliers competing in open markets to provide the world's best electric services, and all of this supported by a new energy infrastructure built on superconductivity, distributed intelligence and resources, clean power, and the hydrogen economy.

The challenges are great, but the opportunities are greater. Please join with the U.S. Department of Energy and the Office of Electric Transmission and Distribution in helping to create a more prosperous, efficient, clean, and secure electricity future for all Americans.

"National leadership is needed to create a shared vision of the future and to build effective public-private partnerships for getting there."

EXECUTIVE SUMMARY

On April 2-3, 2003, 65 senior executives representing the electric utility industry, equipment manufacturers, information technology providers, Federal and state government agencies, interest groups, universities, and National Laboratories met to discuss the future of North America's electric system. (A list of the participants can be found in the appendix.)

The intent of the meeting was to identify a national vision of the future electric system, covering the entire value chain: generation, transmission, distribution, storage, and end-use. The focus was on electric delivery – "the grid," or the portion of the electric infrastructure that lies between the central power plant and the customer – as well as the regulatory framework that governs system planning and market operations.

VISION

"Grid 2030" energizes a competitive North American marketplace for electricity. It connects everyone to abundant, affordable, clean, efficient, and reliable electric power anytime, anywhere. It provides the best and most secure electric services available in the world.

The purpose of this document is to describe the common vision articulated at that meeting. The U.S. Department of Energy will use this vision to help implement President Bush's call for "…modernizing America's electric delivery system" and the 51 recommendations contained in the *National Transmission Grid Study*. Various stakeholders, including industry practitioners, policy makers, and researchers, will use the vision as the coordinating foundation for actions leading to the construction of a 21st century electric system. The vision will guide the development of the National Electric Delivery Technologies Roadmap.

The meeting proceedings, which includes the presentations and summaries of the notes from the facilitated breakout sessions, can be downloaded at www.energetics.com/electric.html.

Major Findings

- America's electric system, "the supreme engineering achievement of the 20th century," is aging, inefficient, and congested, and incapable of meeting the future energy needs of the *Information Economy* without operational changes and substantial capital investment over the next several decades.

⊕ Unprecedented levels of risk and uncertainty about future conditions in the electric industry have raised concerns about the ability of the system to meet future needs. Thousands of megawatts of planned electric capacity additions have been cancelled. Capital investment in new electric transmission and distribution facilities is at an all-time low.

⊕ The regulatory framework governing electric power markets – both at the Federal and state levels – is also under stress. Efforts to loosen regulations and unleash competition have generally fallen short of producing their expected results.

⊕ There are several promising technologies on the horizon that could help modernize and expand the Nation's electric delivery system, relieve transmission congestion, and address other problems in system planning and operations. These include advanced conductors made from new composite materials and high temperature superconducting materials, advanced electric storage systems such as flow batteries or flywheels, distributed intelligence and smart controls, power electronics devices for AC-DC conversion and other purposes, and distributed energy resources including on-site generation and demand management.

⊕ The revolution in information technologies that has transformed other "network" industries in America (e.g., telecommunications) has yet to transform the electric power business. The proliferation of microprocessors has led to needs for greater levels of reliability and power quality. While the transformation process has begun, technological limitations and market barriers hinder further development.

⊕ It is becoming increasingly difficult to site new conventional overhead transmission lines, particularly in urban and suburban areas experiencing the greatest load growth. Resolving this siting dilemma, by a) deploying power electronic solutions that allow more power flow through existing transmission assets and b) developing low impact grid solutions that are respectful of land use concerns, is crucial to meeting the nation's electricity needs.

Conclusions

⊕ The "technology readiness" of critical electric systems needs to be accelerated, particularly for high-temperature superconducting cables and transformers, lower-cost electricity storage devices, standardized architectures and techniques for distributed intelligence and "smart" power systems, and cleaner power generation systems, including nuclear, clean coal, renewable, and distributed energy devices such as combined heat and power.

- A breakthrough is needed to eliminate the "political log jam" and reduce the risks and uncertainties caused by today's regulatory framework. This includes clarifying intergovernmental jurisdiction, establishing "rules of the road" for workable competitive markets wherever they can be established, ensuring mechanisms for universal service and public purpose programs, and supporting a stable business climate that encourages long-term investment.

- Industry will be investing billions of dollars over the next several decades to replace electric power equipment. The economic life of this new equipment will last 40 years or more, so this turnover of the nation's capital stock of electric power assets needs to include the latest technologies to ensure clean, efficient, reliable, secure, and affordable electricity for generations to come. An expanded research, development, and deployment effort is paramount.

- A logical next step is the collaborative development of a *National Electric Delivery Technologies Roadmap*. This technology Roadmap will be principally used for guiding public and private research, development, and demonstration programs.

  - The Roadmap should build on prior similar efforts by the Electric Power Research Institute, National Rural Electric Cooperative Association, North American Electric Reliability Council, California Energy Commission, New York State Energy Research and Development Authority, and others.

  - The Roadmap should complement technology roadmapping efforts by the U.S. Department of Energy in nuclear, renewable, and clean coal power generation systems, energy efficiency technologies, and hydrogen energy systems.

  - The Roadmap should focus on technologies for the electricity delivery portion of the Nation's power grid and should address technology needs for energy assurance and security.

"An expanded research, development, and deployment effort is paramount."

"A logical next step is the collaborative development of a National Electric Delivery Technologies Roadmap...the Roadmap should complement similar efforts ... and focus on electricity delivery..."

TABLE OF CONTENTS

"Grid 2030" — A National Vision for Electricity's Second 100 Years

1 INTRODUCTION

Sixty-five senior executives from electric and gas utilities, equipment manufacturers, information technology companies, Federal and state government agencies, labor unions, interest groups, universities, and National Laboratories participated in the "National Electric System Vision Meeting" held in Washington, D.C., on April 2 and 3, 2003. (A list of the participants can be found in the appendix.) This document reflects the ideas and priorities put forth by the meeting participants.

The meeting was held in response to specific recommendations contained in the Bush Administration's *National Energy Policy*, the *National Transmission Grid Study*, and the *Report of the Secretary of Energy's Electricity Advisory Board*. Together, these documents outline a comprehensive approach to modernizing and expanding America's electricity delivery system. They cover regulatory, market, and technology development strategies. They each call for stronger Federal leadership in overcoming the challenges that currently hinder the development of a "21st century" electric grid.

Photo Removed Due to Copyright Restrictions

America's electric system cannot be modernized and expanded by industry or government alone. This task requires strong public-private partnerships to strengthen the effectiveness of the regulatory framework, foster a stable business climate for long-term investments, and conduct a comprehensive research, development, and demonstration program in advanced electricity technologies.

The U.S. Department of Energy has recently created the Office of Electric Transmission and Distribution to provide stronger leadership and serve as the focal point for policy and technology development activities in the Department related to the electric grid. This Vision, and the forthcoming Roadmap, are critical tools for coordinating activities between the Department, all of the participating organizations, and other stakeholders.

Several efforts by industry have identified future needs and strategies for electricity. Especially notable are the prior efforts performed by the Electric Power Research Institute, the National Rural Electric Cooperative Association, North American Electric Reliability Council, and the California Energy Commission. While these activities have been successful, they are geared toward charting a course of action only for the respective organizations. The U.S. Department of Energy has launched this effort to identify potential actions for all affected stakeholders and to determine where Federal leadership and Federal funding can be best applied to leverage private capital and state and local investments. Unlike other planning activities, this work does not attempt to forecast future conditions and possibilities or outline a range of potential scenarios. It focuses on developing a broad vision, one that is based on the desires and input of a diverse set of stakeholders and can be achieved if all of the goals and actions are met.

The vision and roadmap process has been a useful technique for organizing research, development, and demonstration partnerships involving the U.S. Department of Energy, industry, universities, and National Laboratories. The process typically involves two meetings: 1) a vision meeting attended by senior executives, policy officials, industry leaders, and visionaries; and 2) a subsequent roadmap meeting attended by technology managers, experts, and practitioners. The first establishes "why" and "what." The second addresses "how" and "when."

The diverse set of participants at the vision meeting provided a broad range of perspectives and opinions. The meeting included discussions of the following topics:

- The status of America's electric system today

- The factors – both supporting and inhibiting – affecting the future modernization and expansion of the electric system

- A vision of the future electric system

- The "grand challenges" to be overcome in achieving the future vision

- Key strategic goals that need to be achieved along the way

This document is organized into chapters that correspond with these discussion topics.

"The U.S. Department of Energy has launched this effort to identify potential actions for all affected stakeholders and to determine where Federal leadership and Federal funding can be best applied to leverage private capital and state and local investments."

2 ELECTRIC POWER SYSTEM TODAY

Electric power is essential to modern society. Economic prosperity, national security, and public health and safety cannot be achieved without it. Communities that lack electric power, even for short periods, have trouble meeting basic needs for food, shelter, water, law, and order.

In 1940, 10% of energy consumption in America was used to produce electricity. In 1970, that fraction was 25%. Today it is 40%, showing electricity's growing importance as a source of energy supply. Electricity has the unique ability to convey both energy and information, thus yielding an increasing array of products, services, and applications in factories, offices, homes, campuses, complexes, and communities.

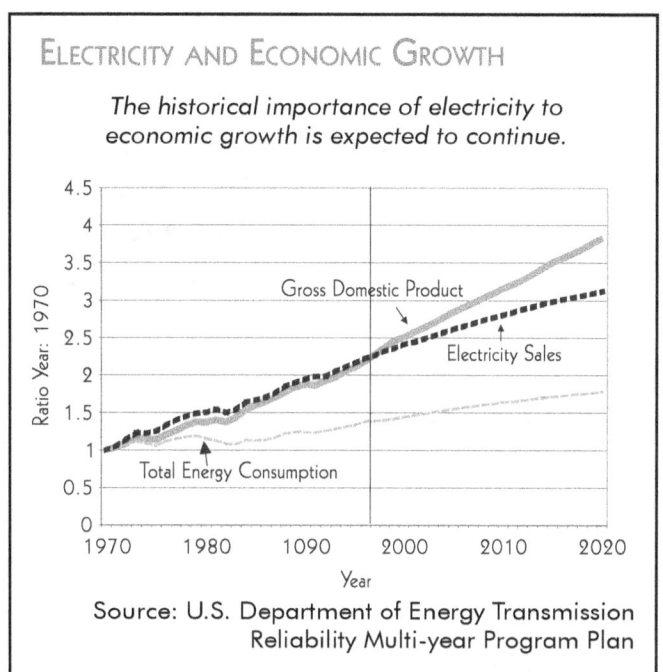

ELECTRICITY AND ECONOMIC GROWTH

The historical importance of electricity to economic growth is expected to continue.

Ratio Year: 1970

Gross Domestic Product

Electricity Sales

Total Energy Consumption

Year: 1970, 1980, 1090, 2000, 2010, 2020

Source: U.S. Department of Energy Transmission Reliability Multi-year Program Plan

The economic significance of electricity is staggering. It is one of the largest and most capital-intensive sectors of the economy. Total asset value is estimated to exceed **$800 billion**, with approximately 60% invested in power plants, 30% in distribution facilities, and 10% in transmission facilities.

Annual electric revenues – the Nation's "electric bill" – are about **$247 billion**, paid by America's **131 million** electricity customers, which includes nearly every business and household. The average price paid is about 7 cents per kilowatt-hour, although prices vary from state to state depending on local regulations, generation costs, and customer mix.

There are more than **3,100** electric utilities:

⊕ 213 stockholder-owned utilities provide power to about 73% of the customers

- 2,000 public utilities run by state and local government agencies provide power to about 15% of the customers

- 930 electric cooperatives provide power to about 12% of the customers

Additionally, there are nearly **2,100** non-utility power producers, including both independent power companies and customer-owned distributed energy facilities.

The bulk power system consists of three independent networks: Eastern Interconnection, Western Interconnection, and the Texas Interconnection. These networks incorporate international connections with Canada and Mexico. Overall reliability planning and coordination is provided by the North American Electric Reliability Council, a voluntary organization formed in 1968 in response to the Northeast blackout of 1965.

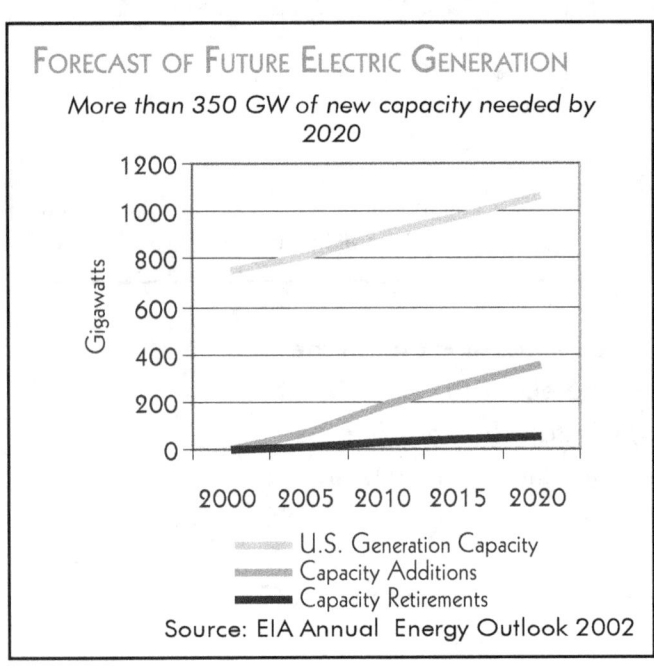

FORECAST OF FUTURE ELECTRIC GENERATION

More than 350 GW of new capacity needed by 2020

Gigawatts

- U.S. Generation Capacity
- Capacity Additions
- Capacity Retirements

Source: EIA Annual Energy Outlook 2002

Electric Generation

America operates a fleet of about 10,000 power plants. The average thermal efficiency is around 33%. Efficiency has not changed much since 1960 because of slow turnover of the capital stock and the inherent inefficiency of central power generation that cannot recycle heat. Power plants are generally long-lived investments; the majority of the existing capacity is 30 or more years old.

The rougly 5,600 distributed energy facilities typically combine heat and power generation and achieve efficiencies of 65% to 90%. Distributed energy facilities accounted for about 6% of U.S. power capacity in 2001.

The profile of the electric power generation industry is changing rapidly, however. A shift in ownership is occurring from regulated utilities to competitive suppliers. The share of installed capacity provided by competitive suppliers has increased from about 10 percent in 1997 to about 35 percent today. Recent data suggest this trend is slowing.

Also, cleaner and more fuel-efficient power generation technologies are becoming available. These include combined cycle combustion turbines, wind energy systems, advanced nuclear power plant designs, clean coal power systems, and distributed energy technologies such as photovoltaics and combined heat and power systems.

Because of the expected near-term retirement of many aging plants in the existing fleet, growth of the information economy, economic growth, and the forecasted growth in electricity demand, America faces a significant need for new electric power generation. In this transition, local market conditions will dictate fuel and technology choices for investment decisions, capital markets will provide the financing, and Federal and state policies will affect siting and permitting. It is an enormous challenge that will require a large commitment of technological, financial, and human resources in the years ahead.

Electric Transmission

Even with adequate electric generation, bottlenecks in the transmission system interfere with the reliable, efficient, and affordable delivery of electric power.

America operates about 157,000 miles of high voltage (>230kV) electric transmission lines. While electricity demand increased by about 25% since 1990, construction of transmission facilities decreased

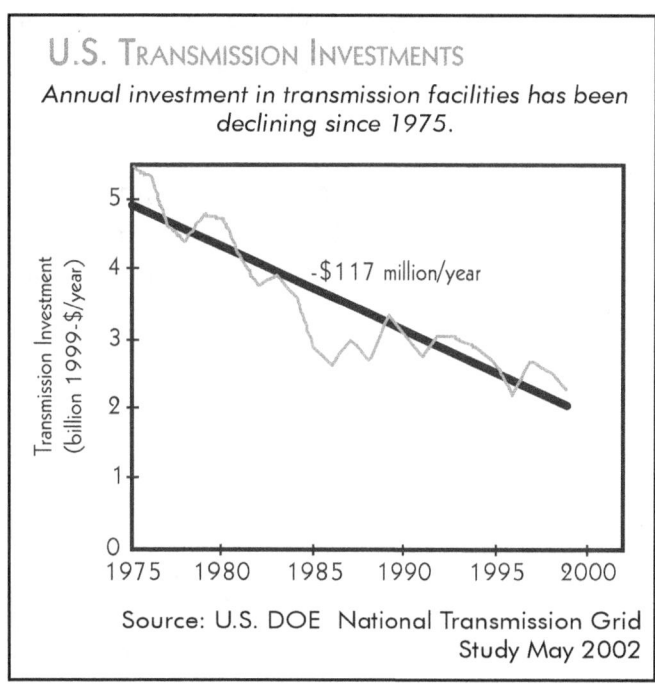

U.S. TRANSMISSION INVESTMENTS

Annual investment in transmission facilities has been declining since 1975.

Source: U.S. DOE National Transmission Grid Study May 2002

about 30%. In fact, annual investment in new transmission facilities has declined over the last 25 years. The result is grid congestion, which can mean higher electricity costs because customers cannot get access to lower-cost electricity supplies, and because of higher line losses. Transmission and distribution losses are related to how heavily the system is loaded. U.S.-wide transmission and distribution losses were about 5% in 1970, and grew to 9.5% in 2001, due to heavier utilization and more frequent congestion. Congested transmission paths, or "bottlenecks," now affect many parts of the grid across the country. In addition, it is estimated that power outages and power quality disturbances cost the economy from $25 to $180 billion annually. These costs could soar if outages or disturbances become more frequent or longer in duration. There are also operational problems in maintaining voltage levels.

America's electric transmission problems are also affected by the new structure of the increasingly competitive bulk power market. Based on a sample of the nation's transmission grid, the number of transactions have been increasing substantially recently. For example, annual transactions on the Tennessee Valley Authority's transmission system numbered less than 20,000 in 1996. They exceed 250,000 today, a volume the system was not originally designed to handle. Actions by transmission operators to curtail transactions for economic reasons and to maintain reliability (according to procedures developed by the North American Electric Reliability Council) grew from about 300 in 1998 to over 1,000 in 2000.

Additionally, significant impediments interfere with solving the country's electric transmission problems. These include: opposition and litigation against the construction of new facilities, uncertainty about cost recovery for investors, confusion over whose responsibility it is to build, and jurisdiction and government agency overlap for siting and permitting. Competing land uses, especially in urban areas, leads to opposition and litigation against new construction facilities.

Electric Distribution

Photo Removed Due to Copyright Restrictions

The "handoff" from electric transmission to electric distribution usually occurs at the substation. America's fleet of substations takes power from transmission-level voltages and distributes it to hundreds of thousands of miles of lower voltage distribution lines. The distribution system is generally considered to begin at the substation and end at the customer's meter. Beyond the meter lies the customer's electric system, which consists of wires, equipment, and appliances – an increased number of which involve computerized controls and electronics which ultimately operate on direct current.

The distribution system supports retail electricity markets. State or local government agencies are heavily involved in the electric distribution business, regulating prices and rates-of-return for shareholder-owned distribution utilities. Also, in 2,000 localities across the country, state and local government agencies operate their own distribution utilities, as do over 900 rural electric cooperative utilities. Virtually all of the distribution systems operate as franchise monopolies as established by state law.

The greatest challenge facing electric distribution is responding to rapidly changing customer needs for electricity. Increased use of information technologies, computers, and consumer electronics has lowered the tolerance for outages, fluctuations in voltages and frequency levels, and other power quality disturbances. In addition, rising interest in distributed generation and electric storage devices is adding new requirements for interconnection and safe operation of electric distribution systems.

Finally, a wide array of information technology is entering the market that could revolutionize the electric distribution business. For example, having the ability to monitor and influence each customer's usage, in real time, could enable distribution operators to better match supply with demand, thus boosting asset utilization, improving service quality, and lowering costs. More complete integration of distributed energy and demand-side management resources into the distribution system could enable customers to implement their own tailored solutions, thus boosting profitability and quality of life.

"The national average load factor is about 55%. This means that electric system assets, on average, are used about half the time."

Demand-Side Management

Customer activities, needs, wants, and desires, as well as the weather, shape patterns of electricity use, which vary by the time of day and season of the year. These patterns typically result in high concentrations of electricity use in "peak periods." The larger the peak period, the greater the amount of electric resources that will be needed to meet it, including distribution, transmission, and generation assets.

The national average load factor (the degree to which physical facilities are being utilized) is about 55%. This means that electric system assets, on average, are used about half the time. As a result, steps taken by customers to reduce their consumption of electricity during peak periods can measurably improve overall electric system efficiency and economics.

Mechanisms to reduce peak demand include time-of-use pricing, load management devices such as "smart" thermostats, load-shifting technologies such as energy storage, and peak-eliminating techniques such as distributed generation and thermally activated heating, cooling, and humidity control devices.

A recent study estimated the potential economic benefits of demand response activities. For example, economic benefits from demand bidding range from about $80 to about $800 million annually, depending on the level of system need. Economic benefits from emergency demand response range from about $85 to more than $300 million annually.[1]

The industry that provides these types of demand-side management and distributed energy products and services is searching for profitable business models. A "boom-bust" cycle is preventing sustainable markets for these businesses from emerging.

Regulatory Framework

America's electric system is vested with the public interest. "Universal electric service" is considered a fundamental part of America's social compact. While complete deregulation of the electric industry is not a realistic goal, restructuring of regulations to open up more segments of the industry to competitive market forces is possible when it is done in a workable manner that increases benefits to customers.

Restructuring of the electricity industry, which began in earnest as a result of the Energy Policy Act of 1992, has been difficult to achieve, for a variety of reasons. For example, the Federal Energy Regulatory Commission regulates interstate wholesale markets. State and local agencies regulate retail markets. The physics of electricity means that markets are typically regional in scope. Yet, the multi-state solutions that are needed for restructuring to occur require a degree of intergovernmental cooperation that has been difficult to achieve.

Uncertainties resulting from the restructuring's bumpy path are interfering with the overall financial health of the industry. Investors are worried about cost recovery and future rates of return for independent power producers, power marketing entities, and investor-owned utilities. These problems have been amplified by the recent financial collapse of several independent power companies and the recent downturn in the economy.

Summary

North America's world-class electric system is facing several serious challenges. Major questions exist about its ability to continue providing

---

[1] U.S. Department of Energy "Report to Congress: Impacts of the Federal Energy Regulatory Commission's Proposal for Standard Market Design" April 2003 DOE/S-0138

citizens and businesses with relatively clean, reliable, and affordable energy services. The recent downturn in the economy masks areas of grid congestion in numerous locations across America. These bottlenecks could interfere with regional economic development. The "information economy" requires a reliable, secure, and affordable electric system to grow and prosper. Unless substantial amounts of capital are invested over the next several decades in new generation, transmission, and distribution facilities, service quality will degrade and costs will go up. These investments will involve new technologies that improve the existing electric system and possibly advanced technologies that could revolutionize the electric grid.

3 FACTORS AFFECTING THE FUTURE OF THE ELECTRIC SYSTEM

Of the many factors shaping future conditions of the electric power industry over the next 20 to 30 years, which will be the most important "drivers" towards a modernized and expanded 21st century electric system?

Public Policy Drivers

Electricity restructuring. One of the most significant public policy drivers is the continuing struggle over the restructuring of the industry. Untangling the "restructuring knot" is one of the premier public policy challenges facing Federal, state, and local energy and environmental policy makers. Many believe that new Federal electricity legislation is required, and for this reason proposals have been submitted in every Congress since 1996, but national consensus has yet to emerge. In the transition from regulated to market operations, the lack of overlap between integrated planning of generation and transmission siting and market-based mechanisms to incent positive investment behavior has exacerbated grid congestion problems and led to poor generation siting decisions. The lack of consensus regarding federal and state jurisdiction in electricity regulations interferes with the expansion and modernization of the grid. Changes in the regulatory framework are needed to create a climate more favorable to risk taking and entrepreneurism.

Environmental regulations. Another significant driver concerns the regulation of the environmental, public health, and safety consequences of electricity production, delivery, and use. This includes

---

THE ELECTRIC POWER RESEARCH INSTITUTE'S ELECTRICITY TECHNOLOGY ROADMAP

One example of another electric system roadmap effort.

Difficult Challenges for Power Delivery and Markets

1. Increasing transmission capacity, grid control, and stability
2. Improving power quality and reliability for precision electricity users
3. Increasing robustness, resilience and security of the energy infrastructure
4. Exploit the strategic value of energy storage
5. Transforming electricity markets
6. Creating the infrastructure for a digital society
7. Electric transportation
8. Technology innovation in electricity use: a cornerstone of economic progress
9. Advances in enabling technology platforms

The roadmap is available at www.epri.com/corporate/discover_epri/roadmap/index.html

---

air pollution, greenhouse gas emissions, land use, and water impacts. Finding ways to address public concerns, reduce the impact of new infrastructure projects, and solve the "not-in-my-backyard" syndrome is critical. Lowering unnecessary costs (time as well as money) associated with overly complex, bureaucratic, and multi-jurisdictional siting and permitting processes is also important. Reconciling legitimate local land use and environmental concerns with the imperative to meet electric deliverability and reliability standards requires the rapid deployment of highly effective, unobtrusive, low-environmental-impact grid technologies.

"The aging of the electric infrastructure... could accelerate turnover of capital assets, including generation, transmission, and distribution facilities."

Historically, environmental requirements have been a source of stimulus to the development of new post-combustion control technologies to meet health and safety standards. However, environmental regulations have neither rewarded nor stimulated new approaches to efficiency. Today, finding ways to streamline environmental requirements (without affecting stringency) may be a more effective means to stimulate stock turnover and the installation of new electricity technologies. Today's air quality regulations are largely being met by the addition of scrubbers and other control technologies on power plants. Greater use of market-oriented environmental policies can provide new revenue streams to enhance the return-on-investment in cleaner and more energy efficient technologies and equipment.

National security. Concerns about national security policies and the need to secure the electric system from threats of terrorism and extreme weather events are affecting the future of America's electric system. A small number of very large generating plants are inherently more vulnerable than a large number of smaller, widely distributed plants. Electric infrastructure and information systems must be secure. Techniques must exist for identifying occurrences, restoring systems quickly after disruptions, and providing services during public emergencies. Policy guidance is needed to clarify roles and responsibilities among electric service providers, regulatory agencies, customers, and law enforcement agencies.

Market Drivers

Competition. The ability of incumbent companies to respond to increasing competition from new entrants is a critical driver determining the future of the electric system. Restructuring of wholesale markets has already created new business opportunities in competitive electric power generation. Restructuring has led many utilities to divest generation assets, agree to mergers and acquisitions, and diversify their product portfolios. Cost cutting measures have included steep reductions in research and development expenditures.

If and when restructuring takes hold, and the array of choices available to customers increases, providers of distributed energy and demand-side management technologies could see the markets for their products and services grow.

Aging infrastructure. Another driver affecting the future of the electric system is the retirement and replacement schedule for generation, transmission, and distribution facilities. The aging of the electric infrastructure coupled with demand increases and more stringent environmental requirements could accelerate turnover of capital assets, including generation, transmission, and distribution facilities. The spread in the use of information technologies has accelerated product cycle times in other sectors, and could do the same in the electric industry.

Consumer demands. As markets become increasingly open to competition, customer wants and desires will play an increasingly stronger role in shaping investment decisions. Concerns about environmental quality, public health, and safety can be seen in customer preferences for renewable energy and energy efficient products and services. Customer interest in affordability, convenience, and on-site control will also drive the design and development of new electric systems. As a consequence, the amount of and demand for consumer information will grow.

Technology Drivers

Information technologies.
Information technologies (IT) have already revolutionized telecommunications, banking, and certain manufacturing industries. Similarly, the electric power system represents an enormous market for

> ### WAMS
>
> The Wide Area Measurement System (WAMS) is a smart, automatic network that applies real time measurements in intelligent, automatic control systems to operate a reliable, efficient, and secure electric transmission infrastructure. WAMS is in place in the West, where it continuously monitors grid performance across the power system. It provides operators with high-quality data and analysis tools to detect impending grid emergencies or to mitigate grid outages. A prototype real time network will be established in the Eastern interconnection, with possibly a dozen measurement units feeding a data collection server by the end on this calendar year. Ultimately, WAMS will monitor the grid parameters in real time, facilitate calculating locational margin prices in real time to support market designs, and assist in providing customer price transparency.

the application of IT to automate various functions such as meter reading, billing, transmission and distribution operations, outage restoration, pricing, and status reporting. The ability to monitor real-time operations and implement automated control algorithms in response to changing system conditions is just beginning to be used in electricity. Distributed intelligence, including "smart" appliances, could drive the co-development of the future architecture for telecommunications and electric power networks, and determine how these systems are operated and controlled. Data access and data management will become increasingly important business functions.

New materials. New scientific discoveries affect America's electric system. For example, advances in the materials sciences are resulting in new conductors of electric power. Nanoscience is opening new frontiers in the design and manufacture of machines at the molecular level for fabricating new classes of metals, ceramics, and biological materials for industrial, computer, and medical applications. Advances in semiconductor-based power electronics have given rise to new solutions that allow more power flow through existing assets, while respecting local land use concerns. Desirable properties of new material for electricity conductors include greater current-carrying capacity, lower electrical resistance, lighter weight, greater controllability, and lower cost.

HTS System Shatters Previous Records

The U.S. Department of Energy and Southwire Company have partnered in a successful high temperature superconductivity (HTS) project near Atlanta, Georgia. The 100-foot, 3-phase HTS power cable system has been in operation since February 18, 2000. It is now operating unsupervised, and Southwire estimates that similar systems will be available on the market in 2005.

Source: Project Fact Sheet,
http://www.eere.energy.gov/superconductivity/

High temperature superconductors. High temperature superconductors are a good example of advanced materials that have the potential to revolutionize electric power delivery in America. The prospect of transmitting large amounts of power through compact underground corridors, even over long distances, with minimal electrical losses and voltage drop, could significantly enhance the overall energy efficiency and reliability of the electric system, while reducing fuel use, air emissions, and physical footprint. Superconducting technologies can be used in generators, cables, transformers, storage devices, synchronous condensors, and motors – equipment that crosscuts the entire electric power value chain.

Electricity storage. Breakthroughs that dramatically reduce the costs of electricity storage systems could drive revolutionary changes in the design and operation of the electric power system. Peak load problems could be reduced, electrical stability could be improved, and power quality disturbances could be eliminated. Storage can be applied at the power plant, in support of the transmission system, at various points in the distribution system, and on particular appliances and equipment on the customer's side of the meter.

Advanced power electronics. High-voltage power electronics allow precise and rapid switching of electrical power. Power electronics are at the heart of the interface between energy storage and the electrical grid. This power conversion interface—necessary to integrate direct current or asynchronous sources with the alternating current grid—is a significant cost component of energy storage systems. Additionally, power electronics are the keystone to Flexible Alternating Current Transmission Systems (FACTS), or power flow controllers, that improve power system control, and help increase power transfer levels. New power electronics advances are needed to lower the costs of these systems, and accelerate their application on the network.

Distributed energy technologies. Developments to improve distributed energy generation and combined heat and power systems could expand the number of installations by industrial, commercial, residential, and community users of electricity. Devices such as fuel cells, reciprocating engines, distributed gas turbines and microturbines can be installed by users to increase their power quality and reliability, and to control their energy costs. They can lead to reduced "upstream" needs for electric generation, transmission, and distribution equipment by reducing peak demand.

Storage Facility Provides Reliable Power for Air Force Base

The U.S. Department of Energy is collaborating with the Tennessee Valley Authority (TVA) to collect and analyze data generated from a new Regenesys Energy Storage System currently under construction in Columbus, Mississippi. When completed, the plant will store 120 MWh of electricity during off peak periods for support of a transmission line

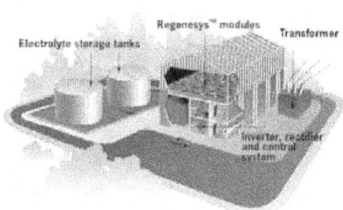

feeding a major Air Force Base. The unit will provide voltage support, supply extra power to cover high summer peaks, and maintain critical loads at the base during emergencies. The plant is expected to begin operation in the spring of 2004. This will be the first application of this technology in the U.S. A similar unit is being constructed in Great Britain.

---

4 VISION OF THE FUTURE ELECTRIC SYSTEM

"Grid 2030" energizes a competitive North American marketplace for electricity. It connects everyone to abundant, affordable, clean, efficient, and reliable electric power anytime, anywhere. It provides the best and most secure electric services available in the world.

This vision of the future electric system builds on the existing electric infrastructure. The same types of equipment that the system uses for electric delivery today - e.g., power lines, substations, and transformers - will continue to play important roles. However, the emergence of new technologies, tools, and techniques including distributed intelligence and distributed energy resources, will increase the efficiency, quality, and security of existing systems and enable the development of a new architecture for the electric grid. The result will be improvements in the efficiency of both power delivery and market operations, and a high-quality network that provides secure sources of electricity to America.

Grid 2030 is a fully automated power delivery network that monitors and controls every customer and node, ensuring a two-way flow of electricity and information between the power plant and the appliance, and all points in between. Its distributed intelligence, coupled with broadband communications and automated control systems, enables real-time market transactions and seamless interfaces among people, buildings, industrial plants, generation facilities, and the electric network.

Technological breakthroughs in superconductivity have made it possible to deliver large amounts of energy over long distances into congested areas unobtrusively, with near-zero voltage drop. New conductor materials enable two to three times the power through existing rights-of-way. Advances in energy storage and demand-side management technologies have virtually eliminated peak-load problems. Economic losses from power outages and power quality disturbances are extremely rare (never caused by electric resource constraints), and customers routinely obtain electricity services at reliability and quality levels tailored to their individual needs with greatly reduced environmental impacts.

"Grid 2030 is a fully automated power delivery network...ensuring a two-way flow of electricity and information between the power plants and appliances and all points in between."

CONCEPTUAL DESIGN OF THE
"GRID 2030" VISION[1]

National Electricity Backbone for Coast-to-Coast Power Exchange

Electricity Backbone Plus Regional Interconnection

Electricity Backbone, Regional Interconnection, Plus Local Distribution, Mini- and Micro-Grids

[1] These are examples for illustrative purposes. The first phase of the Electric Delivery Technologies Roadmap will be to design the architecture of the "Grid 2030" vision.

Workably competitive markets are in place at wholesale levels and customers widely acknowledge the resulting benefits. Effective public oversight and well-designed markets ensure that market power problems are kept to a minimum. Electric transmission and distribution operates under a consistent and stable set of regulations, which rely on performance-based principles and involve Federal and state agencies, multi-state entities, voluntary industry associations, and public interest groups to enforce proper business practices and ensure consumer protection.

The Grid 2030 workforce draws from the Nation's best scientists, engineers, technicians, and business professionals. Workplaces are safe, and workers enjoy rewarding careers in high-paying jobs.

Grid 2030 consists of three major elements:

✧ A national electricity "backbone"

✧ Regional interconnections, which include Canada and Mexico

✧ Local distribution, mini- and micro-grids providing services to customers and obtaining services from generation resources anywhere on the continent

National Electricity Backbone

High-capacity transmission corridors link the east and west coasts, as well as Canada and Mexico. It is possible to balance electric supply and demand on a national basis. This gives customers "continental" access to electricity supplies, no matter where they or their suppliers are located. The national electricity backbone enables expanded distribution of electricity from:

- Efficient generation from a multitude of sources, serving customers in a non-discriminatory manner, and

- A more efficient system that can take advantage of seasonal and regional weather diversity on a national scale, including demand-side management

The backbone system consists of a variety of technologies. These include controllable, very-low-impedance superconducting cables and transformers operating within the synchronous AC environment; high voltage direct current devices forming connections between regions; and other types of advanced electricity conductors, as well as information, communications, and controls technologies for supporting real-time operations and national electricity transactions. Superconducting systems reduce line losses, assure stable voltage, and expand current carrying capacities in dense urbanized areas with a minimal physical footprint. They are seamlessly integrated with high voltage direct current systems and other advanced conductors for transporting electric power over long distances.

"Superconducting systems...are seamlessly integrated with high voltage direct current systems and other advanced conductors for transporting electric power over long distances."

Advanced materials such as high temperature diamond materials could be applied to the transmission, distribution, and control of electricity. Diamond technology could replace silicon and yield revolutionary improvements in current density.

The cryogenc equipment used for achieving superconductivity in electric transmission is available for other purposes, such as the conversion of hydrogen gas into liquid form. Liquid hydrogen is one of the long-distance transport options for the hydrogen economy. With electricity, hydrogen is the second main energy carrier for the economy. Coupling the development of advanced electricity and hydrogen technologies lowers overall infrastructure costs.

Regional Interconnections

The national backbone is connected with two major North American regional interconnections: East and West. Power from the backbone system is distributed over regional networks. Long-distance transmission within these regions is accomplished using upgraded, controllable AC facilities and, in some cases, expanded DC links.

High-capacity DC interties are employed to link adjacent, asynchronous regions. Regional system planning and operations benefit from real-time information on the status of power generation facilities (central-station and distributed) and loads. Expanded use of advanced electricity storage devices address supply-demand imbalances caused by weather conditions and other factors. Markets for bulk power exchanges will operate efficiently with oversight provided by multi-state entities and voluntary industry entities.

Local, Mini- and Micro-Grids

The nation's local distribution systems are connected to the regional networks, and through that to the national electric backbone. Power from distributed energy facilities flows to and from customers and into the regional network, depending on supply and demand conditions. Real-time monitoring and information exchange enables markets to process transactions instantaneously and on a national basis.

Customers have the ability to tailor electricity supplies to suit their individual needs for power, including costs, environmental impacts, and levels of reliability and power quality. Sensors and control systems link appliances and equipment from inside buildings and factories to the electricity distribution system. Advances in distributed power generation systems and hydrogen energy technologies enable the dual use of transportation vehicles for stationary power generation. For example, hydrogen fuel cell powered vehicles provide electricity to the local distribution system when in the garage at home or parking lot at work.

Potential Benefits

There are a number of ways in which Grid 2030 will benefit the American economy, environment, national security, and people.

An expanded and modernized grid will eliminate electric system constraints as an impediment to economic growth. Robust national markets for electric power transactions

Grid-Friendly Appliance Controller

Pacific Northwest National Laboratory has developed a device, based on the gate array chip, commonly found in cell phones, that monitors line frequency, detects dips, and provides the means to trip a load in a graceful manner. Installed in refrigerators, air conditioners, water heaters and various other household appliances, this device would monitor the power grid and turn appliances off for a few seconds to a few minutes in response to power grid overload.

Grid-Friendly Appliance Controller

By triggering appliances to turn on and off at different times, this device could help control power oscillations that occur in different parts of the grid. Grid Friendly™ appliances reduce some of the load on the system to balance supply and demand.

will encourage growth and open avenues for attracting capital to support infrastructure development and investment in new plant and equipment. New business models will emerge for small and large companies in the provision of a wide variety of new products and services for electricity customers, distributors, transmitters, and generators.

More energy efficient transmission and distribution will reduce line losses and lower combustion of fossil fuel and emission of air pollution and greenhouse gases. More economically efficient system operations and the expanded use of demand-side management techniques will reduce the need for spinning reserves, which might also lower environmental impacts. A modernized national electric grid will facilitate the delivery of electricity from renewable technologies such as wind, hydro, and geothermal that have to be located where the resources are located, which is often remote from load centers.

Faster detection of outages, automatic responses to them, and rapid restoration systems will improve the security of the grid, and make the grid less vulnerable to physical attacks from terrorists. Greater integration of information and electric technologies will involve strengthened cyber-security protections. Expanded use of distributed energy resources will provide reliable power to military facilities, police stations, hospitals, and emergency response centers. This will help ensure that "first-responders" have the ability to continue operations even during worst-case conditions. Greater use of distributed generation will lessen the percentage of generated power that must flow through transmission and distribution systems. Higher levels of interconnection with Canada, Mexico, and ultimately other trading partners will strengthen America's ties with these nations and boost security through greater economic cooperation and interdependence.

"Electricity consumers — from factory and business managers to homeowners and small business operators — will have the ability to customize their energy supplies to suit their individual needs."

As a result, Grid 2030 will enable a more prosperous, healthier, and secure quality of life for all Americans. The range of choice in electricity will expand. Electricity consumers — from factory and business managers to homeowners and small business operators — will have the ability to customize their energy supplies to suit their individual needs. More open and workably competitive markets for electricity will help to control cost increases and ensure high quality services.

5 ACHIEVING THE VISION

Progress toward achieving the vision will proceed through three main phases. Phase I involves progress in research, development, and demonstration of advanced technologies. It also includes efforts to clarify and modernize the regulatory framework. Phase II involves turnover of the capital stock of electric assets and replacement with advanced systems. It also involves local and regional deployment of *Grid 2030* concepts and equipment. Phase III involves the extension of local and regional deployment of *Grid 2030* to national and international markets.

A SAMPLE OF POTENTIAL PRODUCTS AND SERVICES MADE POSSIBLE BY PROGRESS TOWARD THE VISION

By 2010	By 2020	By 2030
⊕ Customer "gateway" for the next generation "smart meter", enabling two-way communications and a "transactive" customer-utility interface	⊕ Customer "total energy" systems for power, heating, cooling, and humidity control with "plug&play" abilities, leasable through mortgages	⊕ Highly reliable, secure, digital-grade power for any customer who wants it
⊕ Intelligent homes and appliances linked to the grid	⊕ "Perfect" power quality through automatic corrections for voltage, frequency, and power factor issues	⊕ Access to affordable pollution-free, low-carbon electricity generation produced anywhere in the country
⊕ Programs for customer participation in power markets through demand-side management and distributed generation	⊕ HTS generators, transformers, and cables will make a significant difference	⊕ Affordable energy storage devices available to anyone
⊕ Advanced composite conductors for greater transmission capacity	⊕ Long distance superconducting transmission cables	⊕ Completion of a national (or continental) superconducting backbone
⊕ Regional plans for grid expansion and modernization		

Grand Challenges

A number of grand challenges need to be overcome to achieve this vision of the future electric system. The industry has historically had a monopoly structure that is slow to change with cultural attitudes marked by built-in institutional inertia. Uncertainty about the future has made it difficult for the industry to attract capital investment recently for new construction.

Additionally, technological innovation in the equipment that comprises the electric infrastructure has been somewhat stagnant in recent years.

Overcoming Inertia	Attracting Resources	Developing Better Technologies	Finding Profitable Business Models	Addressing Customer and Public Needs	Developing Better Public Policies
Fragmented industry subject to balkanization	Capital investment	Unobtrusive power lines	Monetizing revenue streams	Demand-side participation in power markets	Federal-state cooperation
Embedded value of capital assets	Education, training, and development of America's workforce	Lower cost storage	Matching rewards to risks	Workably competitive market designs	Market power of incumbent suppliers
Low level of RD&D spending		Long distance superconductivity	Testing Versions 1.0	NIMBY	Public purpose programs
Attitudes resistant to change		Clean power generation			Stable regulatory framework
Slow turn-over of the capital stock		Real-time information systems			Sustained RD&D funding
Lack of success in some markets		Advanced composite conductors			

There is a need to develop and deploy advanced technologies to move the industry from the electro-mechanical to the digital age. Unfortunately, the electric industry is among the lowest in allocating spending to research, development, and demonstration programs.

New entrants into the market (e.g., independent power companies, energy service companies, distributed energy providers, and demand-side management businesses) need the incentives provided by regulatory certainty and properly operating markets in order to realize the necessary revenue streams that will ensure an optimal level of infrastructure investment. Such an environment will sustain profitable business models and lead to the best model that benefits consumers.

The current regulatory framework is not providing a stable business climate. It is not conducive to attracting capital investment, consistent across the country, or consistently effective in addressing public purposes for environmental and consumer protection.

Strategic Goals

Achieving the vision requires progress be made between now and 2030 on a number of important strategic goals.

Transmission	Distribution	Demand-Side Management	Regulatory Framework
Phase I - By 2010			
Prove feasibility of superconducting backbone	Distributed intelligence feasibility proven	Demand-side management programs more widely used	National legislation clarifies jurisdiction issues
Coordinated regional planning and operations	Remote outage detection in place	Smart appliance feasibility proven	Public-private RD&D partnership flourish
Real time information transparency for all grid operators	Plug&play protocols for DG/DR	Greater use of customer side DG/CHP	Workable markets achieved for all sectors and regions
Multiple 10 mile lengths of superconducting cables deployed	Architecture defined for intelligent automated systems		Adequate public subsidies ensure universal service
Advanced planning and initial deployments for the first superconducting "power-hubs" in congested areas	Improved utilization and lower costs		States resolve performance-based regulation, metering, and pricing issues
Smart, automated, grid operation prototype			
Majority of new transmission lines are composite conductors			
Phase II - By 2020			
Half the power in the U.S. flows over smart grid	DG/DR technologies fully integrated in distributed operations	All appliances have smart capabilities	Stable, equitable regulatory framework in place
Long distance superconducting cables installed; "power hubs" operational in several metropolitan areas	Intelligent automated architecture deployed	Large and small customers have access to power markets and real-time information and controls	Workably competitive markets wherever feasible
Average grid losses reduced by 50%	Real-time, two-way flow of information and power		
Phase III - By 2030			
Superconducting backbone installed with fault limiters and transformers	Low-cost, small-scale storage	Fully automated demand response	
Two regional networks	Superconducting cables and equipment deployed	Low cost onsite storage deployed	
100% of power flows through smart grid		Fully interconnected customers and electric networks	

6 PATH FORWARD

For significant progress toward a modernized and expanded electric system to occur, a stronger public-private partnership needs to form and spring into action immediately. Partnership activities aimed at technology development will need to be expanded to address the full set of national needs.

A useful mechanism for guiding these expanded activities is the collaborative development of a *National Electric Delivery Technologies Roadmap*. This roadmapping process will build on existing efforts by the Electric Power Research Institute, National Rural Electric Cooperative Association, Gas Technology Institute, California Energy Commission, New York State Energy Research and Development Authority, and others, including National Laboratories, Universities, and other Federal agencies such as the Department of Defense, National Institute of Standards and Technology, National Science Foundation, Environmental Protection Agency, Tennessee Valley Authority, and the Department of Energy's Power Marketing Administrations.

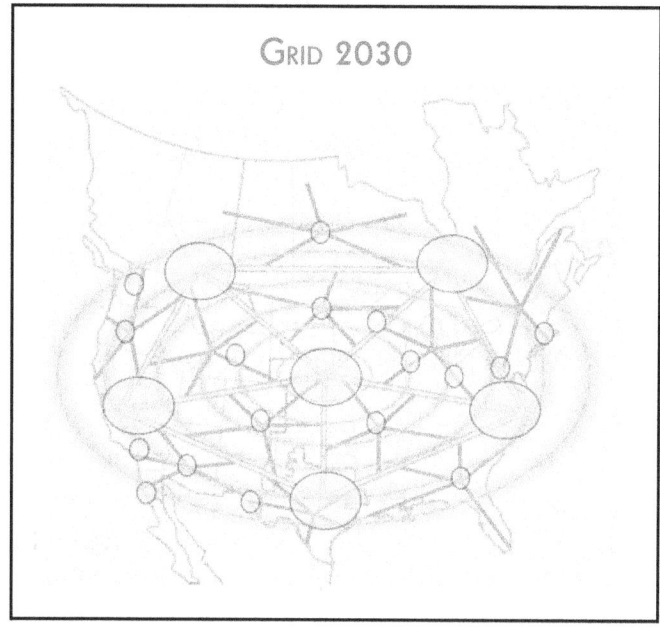

GRID 2030

The *National Electric Delivery Technologies Roadmap* process can be used to identify near-, mid-, and long-term actions and set priorities for research, development, and demonstration programs. It can outline the respective roles of Federal and state government agencies, utilities, equipment manufacturers, trade associations, professional societies, universities, National laboratories, environmental organizations, and other non-governmental organizations. International issues and opportunities can be addressed, including those associated with operating the North American grid with Canada and Mexico, and international trade with partner nations overseas.

The scope of the roadmap process will focus on electricity delivery technologies. Power generation and end-use efficiency technologies have ongoing technology roadmaps and public-private partnerships. Coordination with these complementary efforts will be paramount.

Information from the roadmap could be used by the Federal Energy Regulatory Commission and state public utility commissions to address regulatory framework issues. Specific areas to address in electric delivery technologies include:

- Advanced conductors

- Electric storage

- Sensors and controls

- Distributed intelligence

- Information and communication

- Advanced materials

- Data acquisition, visualization, and simulation modeling

- Advanced power electronics

Working together, a willing coalition of industry, universities, non-governmental organizations, and Federal and state government agencies can help make Grid 2030 a reality. Greater cooperation is needed to address the regulatory framework, capital investment, and technology development. Through this vision, and the subsequent technology roadmap, the U.S. Department of Energy's Office of Electric Transmission and Distribution is eager to assist in a national effort to modernize and expand America's electric delivery system.